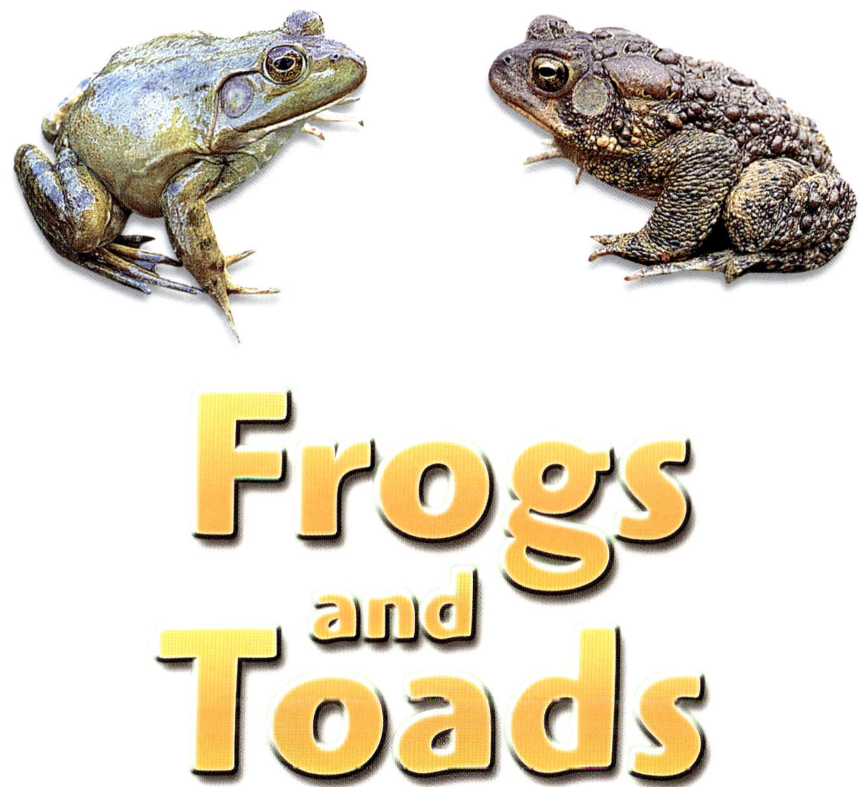

Frogs and Toads

by Ellen Catala

Rigby • Steck-Vaughn

www.HarcourtAchieve.com
1.800.531.5015

Contents

Splash! What was that?
Something just jumped into the water.
Was it a frog or a toad?

Frogs and toads are very much alike.
They are both in a group called **amphibians**.

Amphibians live both in water and on land.
They breathe through their skin when underwater.

Frogs and toads lay their eggs in water.
The babies that hatch are called **tadpoles**.

At first, tadpoles look more like fish.
Then their tails disappear.

Frogs and toads eat only live animals.
They like insects, spiders, worms, and fish.

Frogs and toads take care not to be eaten.
Some frogs have poison to protect themselves.

Frogs have smooth, wet skin.
They are often green but can be other colors.

Toads have dry, bumpy skin.
Many toads are brown or gray.

Frogs have long back legs.
They are very good jumpers.

Toads are wider and flatter than frogs.
They have shorter back legs.
They walk or hop.

Frogs live mostly in or near water.
Pond areas are good homes for frogs.

Toads spend most of their lives on land.
Some even live in deserts, where they
dig into the sand.

Frogs and Toads

Frogs	Both	Toads
smooth, wet skin	amphibians	dry, bumpy skin
long back legs	lay eggs in water	shorter back legs
jump	eat live animals	hop or walk
live in or near water most of the time		live on land most of the time